STARTING OUT

baby RHINOCEROSES

MEG GREVE

CREATIVE EDUCATION • CREATIVE PAPERBACKS

CONT

ENTS

I AM A CALF.

I am a baby rhino.

See the bumps
on my nose?
They will grow
to be horns.

I weighed 100 pounds (45 kg) when I was born.

I stood up right after I was born.
My mom stays close to me. I drink her milk.

I walk and run on my hooves. I am fast!

My mom snorts at lions. She says, "Stay away!"

I love to roll
in mud and
play!

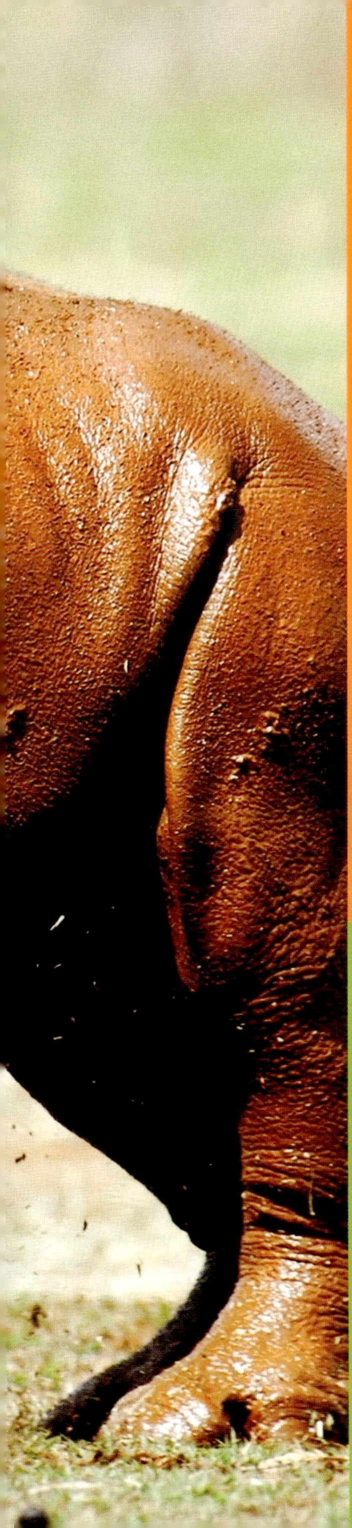

I eat tasty grass.
It helps me grow big.

Someday, I will weigh more than 1,000 pounds (500 kg).

SPEAK AND LISTEN

UHH!

Can you speak like a calf?

Baby rhinos cry and squeal.

Listen to these sounds:

https://www.youtube.com/watch?v=EOoXBOiRDfk

RHINO WORDS

hooves: hard, nail-like coverings on a rhino's feet

horns: hard, sharp parts that grow from a rhino's head

rhino: short for *rhinoceros*; a large land animal with thick, folded skin

snorts: makes a loud blowing noise

READING CORNER

Adamson, Thomas K. *Rhinoceros vs. African Elephant*. Minnetonka, Minn.: Bellwether Media, 2020.

Joubert, Beverly. *The Ultimate Book of African Animals.* Washington, D.C.: National Geographic Kids, 2021.

Kellett, Jenny. *Rhinoceroses: The Ultimate Rhino Book for Kids*. Melbourne, Australia: Bellanova Books, 2023.

INDEX

PUBLISHED BY CREATIVE EDUCATION AND CREATIVE PAPERBACKS
P.O. Box 227, Mankato, Minnesota 56002
Creative Education and Creative Paperbacks are imprints of The Creative Company
www.thecreativecompany.us

LIBRARY OF CONGRESS CATALOGING-IN-PUBLICATION DATA
Names: Greve, Meg, author.
Title: Baby rhinoceroses / Meg Greve.
Description: Mankato, Minnesota : Creative Education and Creative Paperbacks, [2025] | Series: Starting out | Includes bibliographical references and index. | Audience: Ages 4-7 | Audience: Grades K-1 | Summary: "Introduce beginning readers to the exotic world of baby rhinoceroses with this life science starter. Includes photos, a labeled animal diagram, "Make a Noise" section, glossary, and further resources"-- Provided by publisher.
Identifiers: LCCN 2024014698 (print) | LCCN 2024014699 (ebook) | ISBN 9781640264236 (library binding) | ISBN 9781628329568 (paperback) | ISBN 9781640005877 (ebook)
Subjects: LCSH: Rhinoceroses--Juvenile literature. | Rhinoceroses--Infancy--Juvenile literature.
Classification: LCC QL737.U63 G748 2025 (print) | LCC QL737.U63 (ebook) | DDC 599.66/81392--dc23/eng/20240419
LC record available at https://lccn.loc.gov/2024014698
LC ebook record available at https://lccn.loc.gov/2024014699

DESIGN AND PRODUCTION
Design by Rhea Magaro
Production by Beeline Media and Design, Inc.
Art direction by Tom Morgan

PHOTOGRAPHS by Alamy Stock Photo/Danny Lawson, 6; Shutterstock/2021 Photography, 7, Bohbeh, 13, Chad Wright Photography, 10, Corrie Barnard, 8, Daniel-Alvarez, cover, Eric Isselee, 4, 12, 14, Harriet S, 9, sitayi, 2, TashaBubo, 5

Printed in China